3 4028 10784 6025
HARRIS COUNTY PUBLIC LIBRARY

W9-AKC-847

AFRICAN
LEOPARD CUBS

ALPINE IBEX KIDS

LITTLE KIDS FIRST BIG BOOK OF BABY ANIMALS

Maya Myers

NATIONAL GEOGRAPHIC KIDS

WASHINGTON, D.C.

Contents

MONARCH CATERPILLAR

CHAPTER FOUR

FENNEC FOX

CHAPTER FIVE

SUMATRAN ORANGUTANS

CHAPTER SIX

Introduction

Welcome to the world of baby animals! Of course baby animals are cute, but they are also amazing. This book introduces dozens of different kinds of animals in their earliest days. Readers will find out how these babies are born, where they live, what their families are like, how they get their food, and how quickly they learn to do things on their own—all the things that are important to young humans, too.

Chapter One is about the different types of animal babies readers will encounter throughout the book. It answers questions from "What makes a mammal a mammal?" to "Are birds the only kinds of animals that hatch from eggs?" to "What's the difference between reptiles and amphibians?"

Chapter Two takes readers to grasslands around the world. It features babies that hop, strut, scamper, and flutter around the open plains, such as bunnies, ostrich chicks, elephant calves, and butterflies.

Chapter Three goes into the water to meet babies that swim in and fly over oceans, rivers, and lakes. Here, readers will find the world's biggest babies and some of the smallest, along with ducks, sea turtles, alligators, and hippos.

Chapter Four features babies that live on rocky mountainsides and in dry deserts. The conditions in these places may be extreme, but animals like wild goats and prairie dogs are made for places like these.

Chapter Five goes deep into the forest to find babies born on land, in water, in trees, even underground. Among the leaves, readers will meet frogs, orangutans, sloths, and chickadees.

Chapter Six travels to the coldest ends of Earth to find babies that live on snow and ice. Penguins, polar bears, wolves, and musk oxen are some of the fascinating animals to be found in these polar regions.

How to Use This Book

Colorful photographs illustrate each spread and support the text.

Fact boxes give young readers a quick look at an animal's basic biology and family life: what kind of animal it is, what the baby is called, where it lives, how many brothers and sisters it might have, what it eats, and how big it is when it's born.

FACTS

KIND OF ANIMAL
mammal

BABY NAME
kitten

HOME
grasslands, rocky hills, and forests in Africa and Asia

BABIES
one to six at a time; usually three

FOOD
milk, then rodents, birds, gazelles

SIZE AT BIRTH
about the size of a large apple

The black tufts of fur on the tips of caracals' ears may help them hear better.

GRASSLAND BABIES

Caracal

These cute kittens will turn into fierce hunters.

Newborn caracal kittens are helpless. Their eyes don't fully open for about 10 days. The kittens drink their mother's milk and stay snuggled up with her for a few weeks.

Say my name: KARE-uh-kal

Then they come out of their den and play with their brothers and sisters. They run and jump and wrestle. But they aren't just having fun. They are practicing skills that will make them excellent hunters.

For about a year, young caracals follow their mother around. They watch her hunt and learn what to do. Soon they will be able to jump up as high as 10 feet (3 m) in the air to catch birds in flight.

A caracal mother doesn't build her own den. She finds one that's been left empty by another animal.

How high can you jump?

18

19

Interactive questions in each section encourage conversation related to the topic.

Pop-up facts throughout provide added information about the featured animals.

The back of the book offers **parent tips** that include fun activities that relate to baby animals, along with a helpful **glossary.**

CHAPTER 1

Baby Basics

BROWN BEARS

Baby animals can be tiny or huge, furry or bald, slimy or scaly, but they are all pretty adorable! In this chapter, you will learn about the different animal groups found in this book.

Mammal and Bird Babies

Mammals are a group of animals that have hair. Most are born live from their mother. They drink milk before they eat solid food. They use lungs to breathe. Does this sound familiar? That's because people are mammals!

Mammal babies need their parents to take care of them when they are young. They stay close to their mother or father until they learn to find food and keep themselves safe.

WHITE-TAILED DEER

Birds and mammals are endotherms. An endotherm's body can make its own heat. It can keep itself warm even when the air or water around it is cold.

SEA OTTERS

PEEP! Birds are a group of animals that hatch from eggs. Some kinds of baby birds are covered in soft feathers called down. As the chicks get older, smooth adult feathers grow over the down.

Most baby birds stay in their nest until they start to fly. Bird parents bring food to the chirping babies in the nest. When they learn to fly, babies leave the nest to find a meal.

BLACK-CAPPED CHICKADEE

Reptile and Amphibian Babies

Reptiles are a group of animals that includes snakes, turtles, and lizards. They have dry skin covered with scales or bony plates called scutes. Most reptiles hatch from eggs, but some reptile mothers give birth to live young. Most baby reptiles can find food and keep themselves safe as soon as they are born.

OLIVE RIDLEY SEA TURTLE

Reptiles and amphibians are ectotherms. An ectotherm cannot produce its own heat. Its body temperature changes with the temperature around it.

FROG EGGS

TOAD EGGS

Frogs usually lay eggs in clusters. Toads lay eggs in strings.

Amphibians are a group of animals that includes frogs, toads, and salamanders. They have thin, moist skin. They hatch from eggs, often laid in water. When they hatch, they have a tail that helps them swim. Later, they grow legs to walk. The way their bodies change shape is called metamorphosis.

STRAWBERRY POISON DART FROG

AMERICAN ALLIGATORS

Fish, Octopus, and Insect Babies

Many sea creatures also hatch from eggs. Some fish and octopuses lay hundreds—even thousands—of eggs in the water. Other fish keep their eggs inside their bodies and give birth to live young. Most baby fish know how to find food as soon as they are born.

YELLOW BOXFISH

GIANT PACIFIC OCTOPUS EGGS AND HATCHLINGS

MONARCH CATERPILLAR

Insects hatch from eggs, too. They lay their eggs everywhere: in the dirt, on rocks and logs, in water. Some insects, like butterflies, lay their eggs on the plant the babies will eat once they hatch.

MONARCH BUTTERFLY

Some insects go through metamorphosis, kind of like amphibians do. For example, all butterflies and moths hatch as caterpillars, which look very different from adult butterflies and moths.

What did you look like when you were born?

CHAPTER 2

Grassland Babies

Out on grassy plains and meadows, babies large and small are growing, playing, and learning how to survive.

FACTS

KIND OF ANIMAL
mammal

BABY NAME
kitten

HOME
grasslands, rocky hills, and forests in Africa and Asia

BABIES
one to six at a time; usually three

FOOD
milk, then rodents, birds, gazelles

SIZE AT BIRTH
about the size of a large apple

The black tufts of fur on the tips of caracals' ears may help them hear better.

A caracal mother doesn't build her own den. She finds one that's been left empty by another animal.

Caracal

These cute kittens will turn into fierce hunters.

Say my name: KARE-uh-kal

Newborn caracal kittens are helpless. Their eyes don't fully open for about 10 days. The kittens drink their mother's milk and stay snuggled up with her for a few weeks.

Then they come out of their den and play with their brothers and sisters. They run and jump and wrestle. But they aren't just having fun. They are practicing skills that will make them excellent hunters.

For about a year, young caracals follow their mother around. They watch her hunt and learn what to do. Soon they will be able to jump up as high as 10 feet (3 m) in the air to catch birds in flight.

How high can you jump?

Red kangaroo

Kangaroo moms keep their babies VERY close!

FACTS

KIND OF ANIMAL
mammal

BABY NAME
joey

HOME
grasslands and deserts in Australia

BABIES
one at a time

FOOD
milk, then grass and leaves

SIZE AT BIRTH
about the size of a gummy bear

When a baby kangaroo is born, it doesn't look much like a kangaroo. It is tiny, and it doesn't have any fur. The joey crawls up the mother's fur and into the pouch on her belly. Inside the pouch, the joey drinks its mother's milk.

Kangaroos and other animals that carry their babies in a pouch are called marsupials.

A JOEY IN ITS MOTHER'S POUCH

Staying close to Mom keeps a joey safe from hawks and other predators.

After a few months, the joey wiggles out of the pouch and hops around. It eats grass and shrubs. But when it wants milk or feels scared, it climbs back into the pouch. When the joey is eight months old, it is too big to stay in the pouch. But it still pokes its head in to get milk sometimes!

What do you like to carry in your pockets?

There are almost 300 kinds of marsupials. Here are some marsupial babies—and a few marsupial moms—from around the world.

QUOKKAS

KOALAS

OPOSSUM

QUOLL
AGILE WALLABIES
WOMBAT
NUMBAT

Badger cubs are born in underground dens called setts. Setts have tunnels that connect different rooms.

Outside their homes, badgers dig pits to use as toilets.

Eurasian badger

These animals find their friends when they're young.

FACTS

KIND OF ANIMAL
mammal

BABY NAME
cub

HOME
grasslands and woodlands in Europe and Asia

BABIES
up to six at a time; usually three

FOOD
milk, then earthworms, plants, birds, other small animals

SIZE AT BIRTH
about the size of a plum

These badger cubs might look like they're fighting. But don't worry—they're just playing. Badger cubs play to get to know each other.

Badger moms take care of their cubs for two to three months. Then the cubs find a group of badgers to live with. The group might include other badgers from their own family, but it might not.

When badger cubs join a group, they crawl under the adults. They rub against their bellies. This makes the cubs smell like the others in their new group. Now they can find their friends by sniffing!

How do you find your friends at school?

Baby elephants suck their trunks for comfort, just like human babies suck their thumbs.

Elephant calves have baby tusks that fall out and get replaced by adult tusks—just like human kids' teeth.

African elephant

This baby can give itself a shower!

FACTS

KIND OF ANIMAL
mammal

BABY NAME
calf

HOME
grasslands in Africa

BABIES
one at a time; sometimes twins

FOOD
milk, then grass, leaves, fruits, flowers, roots

SIZE AT BIRTH
about the size of a grocery cart

Elephant calves are the biggest babies on land. But they aren't too big to stay close to Mom. A calf stays with its mother until it's about eight years old. It uses its trunk to hold on to its mother's trunk or tail.

Calves also use their trunks to breathe, drink, smell, and pick up food. They can suck up water and mud with their trunks and spray it into the air. A little shower helps them get clean or cool off.

A baby elephant can take a nap standing up!

Ostrich

The world's biggest chick doesn't sit still for long.

Like their parents, ostrich chicks cannot fly. They use their wings to help them balance when they run.

FACTS

KIND OF ANIMAL
bird

BABY NAME
chick

HOME
grasslands and woodlands in Africa

EGGS
around 10 at a time

FOOD
roots, seeds, flowers, leaves, insects, small reptiles, rodents

SIZE AT BIRTH
as big as an adult chicken

Tap, tap, tap! Ostrich chicks peck their way out of giant eggs. Their mother and father bring them food at first. But in just a few days, the chicks walk out of the nest. They follow their parents and look for seeds and leaves to eat. When they get too hot, they huddle under Mom and Dad for shade.

The ostrich is the biggest bird in the world, and it lays the biggest eggs.

Ostrich chicks eat their parents' poop. It helps them digest their food.

About a month after they are born, the chicks can run—fast! A young ostrich can run as fast as a car drives on a city street. An ostrich's speed helps it get away from predators like cheetahs and lions.

A baby rhino makes high-pitched squealing noises to call its mother. The mom pants, or breathes heavily, in different patterns to call back.

A rhino calf stays close to its mother for about four years. When the mom has a new baby, she chases the older calf away.

Black rhinoceros

This baby is not as tough as it looks.

FACTS

KIND OF ANIMAL
mammal

BABY NAME
calf

HOME
grasslands, forests, and deserts in Africa

BABIES
one at a time; sometimes twins

FOOD
milk, then grass and shrubs

SIZE AT BIRTH
about the size of a bed pillow

The thick skin of a rhinoceros looks a lot like a suit of armor. But because there's no hair to protect it, it is very sensitive. A baby rhino must learn how to take care of its skin in the hot sun.

A calf follows its mother to the watering hole for a swim. The baby rhino rolls in the mud to cool off, just like Mom. The mud sticks to its skin. It helps protect the rhino from sunburn and bug bites.

How do you like to cool off when you're hot?

Monarchs are poisonous to many animals. Their bright orange color warns predators, like birds and other insects, to stay away.

MONARCH EGG

A butterfly's life cycle has four stages: egg, caterpillar, chrysalis, and adult. This is called complete metamorphosis.

Monarch butterfly

Monarch moms fly great distances to lay their eggs.

FACTS

KIND OF ANIMAL
insect

BABY NAME
larva or caterpillar, pupa

HOME
milkweed plants in Asia, Australia, and North America

EGGS
up to 500 eggs total

FOOD
milkweed leaves, then nectar

SIZE AT BIRTH
about the size of a strawberry seed

Monarch butterflies fly hundreds of miles to find the plants their babies will need to eat. They lay one egg at a time on the bottom of milkweed leaves.

About four days later, the tiny egg hatches. Out comes a tiny caterpillar. It begins to munch, munch, munch on the milkweed leaves.

JUST HATCHED MONARCH CATERPILLAR

A butterfly chrysalis is also called a pupa.
MONARCH CATERPILLAR FORMING A CHRYSALIS

The caterpillar eats and eats. It grows bigger and bigger. When the caterpillar is about as long as your finger, it stops eating. It hangs from a stem or branch and makes a hard, thin shell around its body. This shell is called a chrysalis. Inside the chrysalis, an amazing thing happens. The caterpillar changes into a butterfly!

MONARCH BUTTERFLY ABOUT TO EMERGE FROM CHRYSALIS

A MONARCH BUTTERFLY OUT OF THE CHRYSALIS

After about two weeks, the chrysalis splits open. Out comes the new butterfly. Its wings look crumpled and wet. But soon they straighten out and stiffen. Now the butterfly spreads its wings. Fly, butterfly!

Where can you look for butterflies near your home?

Black-backed jackal

Taking care of these pups is a job for the whole family.

Newborn jackal pups drink their mother's milk in their underground burrow. But Mom is not the only one taking care of these babies. The father jackal also works to keep them safe. He guards the den and brings meat for the pups.

FACTS

KIND OF ANIMAL
mammal

BABY NAME
pup

HOME
parts of Africa

BABIES
two to seven

FOOD
milk and meat chewed by adults, then small animals, insects, fruits, nuts

SIZE AT BIRTH
about the size of a large apple

The pups stay with their family for about a year.

Jackal pups start hunting on their own when they are about six months old.

When the pups are about three months old, they come out of the den. They play and learn how to hunt. In this close family pack, Mom and Dad have helpers. Big brothers and sisters babysit their younger siblings. They also bring food for the pups and keep them safe from hungry wolves, leopards, and hyenas.

How could you help take care of a baby?

Rabbit kittens only drink milk from their mother twice a day, at sunrise and sunset.

Rabbits make two kinds of poop. They eat one kind to get extra nutrition.

Eastern cottontail rabbit

Bunnies double their size in the first 10 days of life.

FACTS

KIND OF ANIMAL
mammal

BABY NAME
kitten

HOME
grasslands and shrubby areas in North and South America

BABIES
usually five, up to 12 at a time

FOOD
milk, then grass, clover, vegetables, twigs

SIZE AT BIRTH
about the size of a kiwifruit

In tall grass, a rabbit mother digs a shallow hole for her nest. She lines the nest with leaves, grass, and some of her fur to make a soft place for her babies. The kittens are tiny and furless when they are born. Within a week, they have soft fur.

After two weeks, the baby cottontails are ready to leave the nest. They hop around to find grass to nibble. They play with their brothers and sisters. After another month or so, the siblings hop off on their own.

How far can you hop?

CHAPTER 3

Water Babies

Swim, splash, dive! These babies wiggle and wade in water around the world.

FACTS

KIND OF ANIMAL
reptile

BABY NAME
hatchling

HOME
open ocean and coastal areas of Africa, Asia, and North and South America

BABIES
about 100 at a time

FOOD
jellyfish, crabs, snails, shrimp, kelp, algae

SIZE AT BIRTH
about the size of an apricot

EGGS

Thousands of turtles can lay eggs on the same beach at the same time. This is called an arribada — Spanish for "arrival."

Olive ridley sea turtle

These babies take a dangerous trip as soon as they're born.

Olive ridley turtles are named for the dusty green color of their shells. Hatchlings are black and change color as they grow.

In the dark of night, olive ridley sea turtle moms swim to a beach. They dig nests in the sand and lay their eggs inside. They cover the eggs with sand and go back out to sea.

In two months, the eggs hatch. The hatchlings use their flippers to scurry to the ocean. Many are snatched up by hungry birds and ghost crabs. The hatchlings that make it to the water swim out and dive to find food.

Male olive ridleys stay in the ocean their entire lives. Females return to the beach where they were born to lay their eggs.

Why do you think a sea turtle mom lays so many eggs?

White's seahorse

These babies grow inside their dads.

Did you know that a seahorse dad can have babies? The mom puts her eggs into the dad's special pouch, called a brood pouch.

After three weeks, the eggs hatch. The babies leave the pouch and swim off. They stay close to their brothers and sisters.

FACTS

KIND OF ANIMAL
fish

BABY NAME
fry

HOME
coastal waters near Australia

EGGS
up to 250 at a time

FOOD
plankton and tiny fish or shrimp

SIZE AT BIRTH
about the size of a raisin

Like many other fish, seahorse fry don't need their parents to take care of them. They can find food on their own. Baby seahorses eat and eat all day. They don't have teeth to chew with—they just swallow their food whole.

Seahorses can change color quickly to blend in with their surroundings.

BABY SEAHORSE

Seahorses can hold on to things with their tails. They also hold tails with each other.

Who do you like to hold hands with?

FACTS

KIND OF ANIMAL
bird

BABY NAME
chick

HOME
wetlands in Africa, Asia, and Europe

EGGS
one at a time

FOOD
crop milk, then small crustaceans and algae

SIZE AT BIRTH
about the size of a peach

Greater flamingo

These chicks don't look much like their parents.

A crèche of flamingos can include 3,000 birds!

A fluffy, gray flamingo chick hatches on top of its mud nest. Its mom and dad feed it something most birds don't have: a kind of milk called crop milk. This red fluid comes out of the parents' bills.

When the chick is about a week old, it joins other chicks in a group called a crèche. A few adults look after all the babies in the large flock while the other parents look for food.

A chick's parents still feed their baby when it's in the crèche. Even with thousands of birds in a flock, parents recognize their own chick's calls and find it quickly.

Flamingo parents make a volcano-shaped nest out of mud.

When it's two or three months old, the young flamingo starts feeding itself. Now it can eat the food that will make its feathers—and its legs!—change to a pinkish color. The pink comes from the small creatures and algae that flamingos eat.

It takes two to three years for a flamingo to turn pink.

FACTS

KIND OF ANIMAL
cephalopod

BABY NAME
hatchling

HOME
Pacific Ocean

EGGS
up to 100,000 at a time

FOOD
shrimp, lobsters, fish, clams, birds, sharks

SIZE AT BIRTH
about as big as a grain of rice

The biggest giant Pacific octopus ever found weighed as much as three baby Asian elephants.

Giant Pacific octopus hatchlings have 14 suckers on each arm. Adults have up to 280 suckers on each arm.

Giant Pacific octopus

The biggest octopus in the world hatches from a tiny egg.

Octopus hatchlings have clear bodies that can change color instantly when they need to hide.

In an ocean cave, a giant Pacific octopus mom lays thousands of eggs. She hangs clusters of them from the cave's ceiling. She guards them constantly for about six months.

When the eggs hatch, the mother squirts water to blow the hatchlings out into the ocean. The hatchlings float near the surface for several months, eating tiny creatures. Then they swim down deeper and start to grow—and they never stop. A giant Pacific octopus gets bigger all its life.

How big do you think you will grow up to be?

Sea otter

These babies are made to float.

Sea otters have pockets! While hunting, an otter tucks food into flaps of skin in its armpits.

Sea otter babies are born in the ocean, but they don't learn to swim until they are about a month old. How do these pups survive? They float! Their thick fur traps so much air that they bob on the surface. But otter moms hold their babies on their bellies most of the time.

MOM FEEDS A MUSSEL TO HER PUP.

A sea otter pup stays with its mom for six to eight months.

FACTS

KIND OF ANIMAL
mammal

BABY NAME
pup

HOME
Pacific Ocean

BABIES
one at a time

FOOD
milk, then sea urchins, shellfish, crabs

SIZE AT BIRTH
about the size of a large house cat

Sea otters have the thickest fur of all animals.

The pup eats the sea creatures Mom brings it. To keep her baby from floating away while she dives for food, the mom wraps her pup in seaweed.

What helps you float in the water?

Mallard ducklings snuggle in the nest their mother made.

FACTS

KIND OF ANIMAL
bird

BABY NAME
duckling

HOME
wetlands in Africa, Asia, Australia, Europe, and North America

EGGS
up to 13 at a time

FOOD
worms, insects, seeds, grass

SIZE AT BIRTH
about the size of a tennis ball

Mallard duck

Make way for these busy little birds.

MALLARD EGGS

A day after they hatch, mallard ducklings are ready to leave the nest. Their mom leads them to the water. The ducklings follow her in a line, then jump right in and start swimming. They stick their heads underwater to find food.

A mother mallard makes her nest in tall grass near the water.

When Mom gets out of the water, all the ducklings follow. The ducklings snuggle up with Mom. They will keep very close to her for about two months. Once they can fly, they are ready to be on their own.

Only female mallard ducks quack.

This alligator hatchling is breaking out of its shell. Hello, little gator!

American alligator

These babies go for a ride in their mother's mouth.

FACTS

KIND OF ANIMAL
reptile

BABY NAME
hatchling

HOME
swamps and wetlands in North America

EGGS
about 40 to 50 at a time

FOOD
fish, insects, reptiles, birds

SIZE AT BIRTH
about as long as a banana

In a muddy nest, baby alligators call to their mother from inside their tough shells. They have a special tooth to help them break out. If an egg doesn't open, Mom breaks the shell with her mouth. Once the babies have hatched, their mom gently carries them to the water in her jaws. Her sharp teeth don't hurt the babies.

A group of newly hatched alligators is called a pod.

The baby alligators can swim and find their own food, but they will stay close to Mom for about a year. She keeps away predators like raccoons, bears, otters, and herons—and she gives the babies rides on her snout, too!

What are some other ways animals carry their babies?

Hippopotamus

This baby and its mother are very close.

FACTS

KIND OF ANIMAL
mammal

BABY NAME
calf

HOME
lakes and rivers in Africa

BABIES
one at a time

FOOD
milk, then grass, water plants, fruits

SIZE AT BIRTH
about as long as a kid's bicycle

When it's time for her baby to be born, a mother hippopotamus moves away from her herd. She wants to be alone with her baby.

Hippo calves are often born underwater. The mom nudges the baby to the surface to breathe. The baby dips underwater to drink its mother's milk. The calf closes its ears and nostrils to keep the water out while it drinks. Staying in the water helps keep the calf safe from predators like hyenas and lions.

A hippo's skin oozes an oily red liquid that protects it from sunburn.

A hippo calf can't swim like its mother. It walks or gallops underwater.

Hippos can sleep underwater. They come to the surface to breathe without waking up.

For two weeks, the mom and baby stay together by themselves. They nuzzle and cuddle with each other. Then the mother brings the calf back to the herd. The baby hippo starts playing with other calves.

At night, the calf follows its mother to look for grass to eat. It will stay close to Mom for about seven years, even after a new brother or sister is born.

Only baby boxfish are bright yellow. As they grow, females turn a darker yellow and males turn blue-gray.

When boxfish are in danger, they release poison into the water.

Yellow boxfish

This fish gets its name from its boxy shape.

In the shelter of a coral reef, a yellow boxfish mother lays her eggs in the water. Then she swims away. After the eggs hatch, the hatchlings come to rest on the coral.

As the baby boxfish grow, they swim around and look for food. Their tiny fins are perfect for steering their blocky bodies through the twists and turns of the coral. But these fish stick close to the reef. They are not good swimmers in open water, where the current could sweep them away.

Under its skin, an adult boxfish has a set of hard, bony plates.

FACTS

KIND OF ANIMAL
fish

BABY NAME
hatchling, larva

HOME
oceans off Africa, Australia, and Asia

EGGS
unknown

FOOD
algae and small sea animals

SIZE AT BIRTH
unknown

Can you name some other things that are box-shaped?

ADULT FEMALE BOXFISH

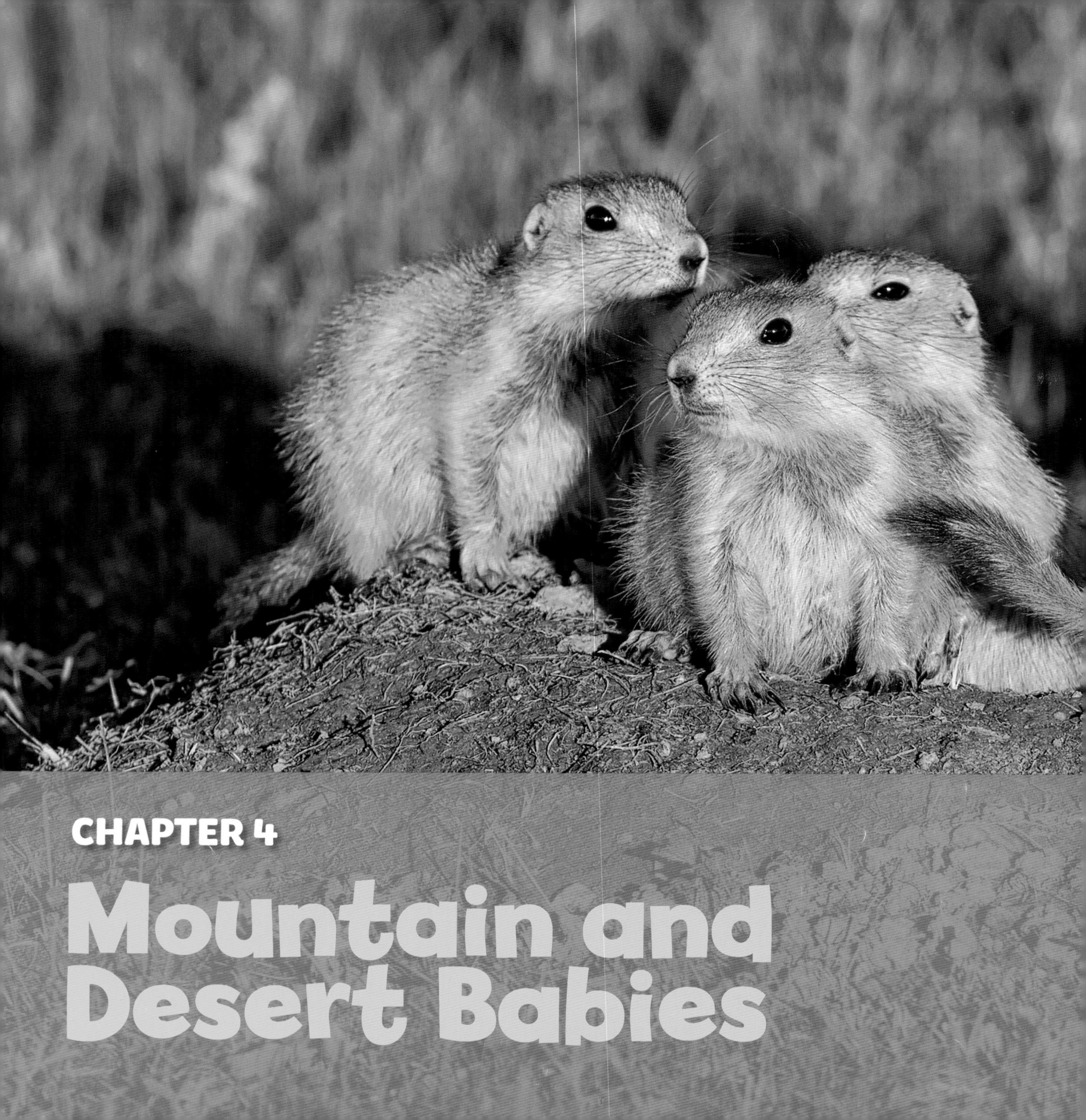

CHAPTER 4

Mountain and Desert Babies

Skills like climbing and digging help baby animals survive on a steep mountainside or in a dry desert.

Brown bear cubs stay with their mother for two or three years.

Brown bear

Cubs follow Mom to find the tastiest treats.

A brown bear mother rests through the winter in her den. Her babies are born inside this cozy home. The new cubs are blind and hairless. For a few months, they drink their mother's milk. Their fur comes in, and they get bigger.

FACTS

KIND OF ANIMAL
mammal

BABY NAME
cub

HOME
mountain forests, meadows, coastlines in Asia, Europe, and North America

BABIES
usually two or three at a time

FOOD
milk, then fruits, fish, honey, roots, nuts, insects, small mammals

SIZE AT BIRTH
about the size of a loaf of bread

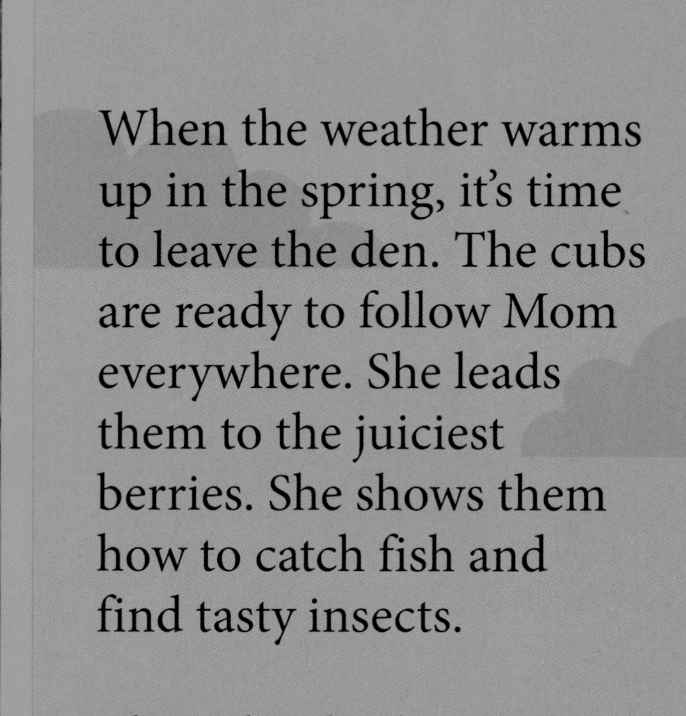

When the weather warms up in the spring, it's time to leave the den. The cubs are ready to follow Mom everywhere. She leads them to the juiciest berries. She shows them how to catch fish and find tasty insects.

The cubs climb trees by themselves to find sticky, sweet honey. Their mom waits on the ground.

Brown bears have an excellent sense of smell.

When they are not looking for food, bear cubs like to play together. They wrestle and pretend to fight. This is how they learn to defend themselves.

When brown bears are very full, they may dig a hole for their bellies to fit into when they lie down.

Where could you look for berries?

Alpine ibex

These wild goats are excellent climbers soon after they're born.

It doesn't take long for ibex kids to get their footing in the mountains. Right after they're born, they can walk and jump. In a few weeks, they follow their mothers onto steep, rocky cliffs.

FACTS

KIND OF ANIMAL
mammal

BABY NAME
kid

HOME
rocky mountainsides in Europe

BABIES
one at a time; sometimes twins

FOOD
milk, then grass, leaves, wood, bark

SIZE AT BIRTH
about the size of a house cat

A mother uses her nose to find her kid in the herd.

Kids live in herds with their mothers and other adult females until they are about a year old.

You can count the ridges on an ibex's horn to find out how old it is.

Ibex kids like to play. They run and jump and push each other off rocks. They pretend to fight by butting their heads together. If a kid wanders too far from its mom, she calls out with a *baa.* The kid calls back with a *baa,* too.

How old do you think this ibex is? Count its ridges.

Golden eagle nests are so big that you and a friend or two could probably fit inside!

A young golden eagle gets its adult feathers when it's about four years old.

Golden eagle

These balls of fluff will become huge birds of prey.

FACTS

KIND OF ANIMAL
bird

BABY NAME
eaglet, hatchling

HOME
mountain forests and wetlands in Africa, Asia, Europe, and northern North America

EGGS
usually two, up to four at a time

FOOD
rabbits, squirrels, prairie dogs, birds, fish

SIZE AT BIRTH
about the size of a peach

Golden eagle parents build their giant nest on a cliff, in a tree, or on the ground. In the nest, pink, scrawny eaglets hatch from speckled eggs. They have grayish white, fluffy down. Soon, some black feathers grow in.

A mother and father eagle spend four to six weeks getting their nest ready for chicks.

The mom stays with the eaglets, and the dad hunts for food. He brings it back to the nest. The eaglets can't fly, but they hop, walk, and use their wings to balance.

When the baby eagles are about 10 weeks old, they learn to fly. Now they're called fledglings. They start hunting for food on their own.

If you could build a nest to live in, what would it look like?

Prairie dogs aren't dogs. They are related to squirrels.

Thousands of prairie dogs can live in one town.

A PRAIRIE DOG FAMILY

Black-tailed prairie dog

These small mammals live in huge towns.

Prairie dog pups live with their families in underground colonies called towns. These towns are made of many rooms. The rooms are connected by tunnels. There are special rooms for sleeping, having babies, and even going to the bathroom!

Pups nestle together in a room called a nursery. They drink their mother's milk here. When they are about six weeks old, the pups leave the nursery. They scurry through the tunnels to head outside for the first time.

FACTS

KIND OF ANIMAL
mammal

BABY NAME
pup

HOME
underground colonies in prairies in North America

BABIES
two to eight at a time

FOOD
milk, then leaves, grass, seeds

SIZE AT BIRTH
about as big as a small plum

Prairie dogs use a special song and dance called a jump-yip to announce it's safe to be outside.

Fennec fox

This small fox's giant ears can hear prey under the sand.

Fennec kits snuggle in their deep, sandy den with their mother. They drink her milk. Their dad brings food for their mom. At first, a kit's ears are small and folded forward, but they grow bigger and bigger.

FACTS

KIND OF ANIMAL
mammal

BABY NAME
kit

HOME
deserts in Africa

BABIES
usually two to four, up to six at a time

FOOD
milk, then insects, rodents, reptiles, fruits, leaves, roots

SIZE AT BIRTH
about the size of a peach

When they are a month old, the kits go outside. Now they have room to run and jump. They play chase with their parents and siblings.

Fur on the bottom of a fennec's feet keeps its paws from burning on the hot sand.

A fennec's giant ears help it stay cool by releasing heat from its body.

These desert foxes can go for a long time without water.

The fennec family spends most of the hot daytime in their den. They come out at night to find food. Fennec foxes listen for small creatures moving through the sand, then they pounce.

What can you hear when you go outside?

These owls can make a rattlesnake-like sound to scare away predators.

Burrowing owl

These chicks roll in the dirt to get clean.

FACTS

KIND OF ANIMAL
bird

BABY NAME
chick, owlet

HOME
underground burrows in North and South America

EGGS
up to 12 at a time

FOOD
grasshoppers, moths, beetles, reptiles, rodents, birds

SIZE AT BIRTH
about the size of a strawberry

Burrowing owls make their nests in underground tunnels. An old burrow dug by a prairie dog or a tortoise makes a good home for these fluffy chicks.

The mother owl stays with the newborn chicks, and the father brings food. When the chicks are two weeks old, the mother starts hunting, too. The chicks wait near the burrow's entrance for their meals to be delivered.

Young owls pretend to hunt by jumping on their food—and on each other! When they are about six weeks old, the owlets leave the burrow to hunt for themselves.

Burrowing owl parents drop food right into their babies' beaks.

CHAPTER 5

Forest Babies

WHITE-TAILED DEER

In the woods, babies are born in trees, on the forest floor, and underground.

A panda cub's cry sounds a lot like a human baby's cry.

A newborn panda cub drinks milk up to 14 times each day.

FACTS

KIND OF ANIMAL
mammal

BABY NAME
cub

HOME
mountain forests in China

BABIES
one at a time

FOOD
milk, then mostly bamboo leaves and shoots

SIZE AT BIRTH
about the size of an orange

Giant panda

A newborn cub is smaller than its mother's ear.

In a cave or a hollow tree, a mother panda gives birth to a tiny cub. The cub has pink skin. It soon grows fuzzy white fur. In about a month, its fur shows a black-and-white pattern.

The mom keeps her baby close. She carries it in her paw or in her mouth. When the cub is a few months old, it can crawl and take wobbly steps. At a year, it can run and climb trees.

Panda cubs drink milk at first. Then they start eating bamboo. They spend most of each day chewing these leaves and woody shoots.

A mother panda may wake her baby up from a nap because she wants to play with it.

If you had to eat only one food most of the time, what would it be?

Black-capped chickadee

This small songbird is easy to recognize by its black head.

A chickadee's extra-long call can be a warning that predators are near.

Chickadee parents make their nest in the soft wood of a dead tree. They line the nest with feathers, moss, and animal hair. The mother sits on her eggs for two weeks, while the dad brings her food.

FACTS

KIND OF ANIMAL
bird

BABY NAME
chick

HOME
forests and bushes in North America

EGGS
usually six to eight, up to 13 at a time

FOOD
seeds, berries, insects, spiders, worms

SIZE AT BIRTH
about the size of a grape

This bird's name sounds like its call: *chick-a-dee-dee-dee.*

Chickadees hide bits of food to eat later. They can remember thousands of hiding places!

Two weeks after the eggs hatch, the tiny chicks are fluffy with feathers. They are ready to fledge, or fly. The family leaves the nest. They stay together as they move around.

The fledglings chirp when they are hungry. For a few more weeks, Mom and Dad keep bringing them caterpillars to eat. The little birds swallow them whole.

Can you make up your own birdcall?

One sloth can have almost a thousand moths and beetles living in its fur!

Two-toed sloth

This clingy baby doesn't want to let go of its mom.

FACTS

KIND OF ANIMAL
mammal

BABY NAME
baby sloth

HOME
tropical forests in South America

BABIES
one at a time

FOOD
milk, then fruits, leaves

SIZE AT BIRTH
about as long as a loaf of bread

Sloths can do just about anything while hanging from a branch. They eat, sleep, and even have babies while hanging upside down in trees! As soon as a baby sloth is born, it clings to its mother's fur and drinks her milk.

The mother sloth carries her baby everywhere for six to nine months. The baby learns where to find the tastiest leaves and juiciest fruits.

Baby sloths have short fur when they're born. It takes several months for their long adult fur to grow in.

Sloths are good at swimming but not at walking. On the ground, they drag themselves along on their bellies.

A baby sloth's long, curved claws help it hold on tight to its mother and then to trees when it gets older.

The baby sloth also learns the one reason sloths go down to the ground. Once a week, the mom carries her baby down the tree to poop on the forest floor. Then they crawl back up into the leaves.

Sloths are usually nocturnal. They sleep during the day and find food at night.

When a sloth is a year or two old, it's time for it to move out of Mom's tree. But it won't go far—sometimes just to the tree next door. It may stay there for the rest of its life.

What is something you do only once a week?

FACTS

KIND OF ANIMAL
amphibian

BABY NAME
tadpole

HOME
tropical rainforests in southern North America

EGGS
three to five at a time

FOOD
special eggs laid by the mother, then insects

SIZE AT BIRTH
about the size of a tomato seed

TADPOLE

The insects they eat make these frogs poisonous to predators.

Poison dart frog tadpoles are brown, but they will grow up to be colorful frogs!

Strawberry poison dart frog

This amphibian rides piggyback on Mom.

TADPOLE IN A BROMELIAD

A strawberry poison dart frog lays her eggs on the forest floor. When they hatch, her tadpoles need water. She carries them on her back and climbs up high to a plant called a bromeliad. This cup-shaped plant fills with little pools of water when it rains. Each tadpole goes into its own pool.

Every day, the mother frog visits the pools and lays a special kind of egg for the tadpoles to eat. The tadpoles eat the eggs.

If the frog mom puts more than one tadpole in each pool, the biggest one will eat the smaller ones.

When a tadpole is about as long as a grain of rice, it grows legs. Its tail disappears. Now it is a frog. It leaves its little pool to explore the forest.

Leopard

These animals move from den to den to keep their cubs safe.

Leopard cubs stalk, pounce, and chase their siblings to practice their hunting skills.

Leopard cubs stay with their mother for one to two years. Then they usually live alone.

A leopard keeps her small cubs hidden in a cave or a hollow tree. She leaves them to hunt, but she moves them to a different hiding spot often to keep predators from discovering them. She picks up the cubs by the back of their necks and carries them in her mouth.

After six to eight weeks, the cubs are old enough to follow their mom around. She shows them how to hunt.

FACTS

KIND OF ANIMAL
mammal

BABY NAME
cub

HOME
forests and grasslands in Africa and Asia

BABIES
two or three at a time

FOOD
milk, then birds, antelope, rodents, monkeys, fish

SIZE AT BIRTH
about the size of a hot dog bun

Leopards are good at climbing trees.

How did your parents carry you when you were little?

Whether they're called kittens or cubs, these young wild cats can definitely be called cute!

TIGER CUB

SAND CAT KITTEN

LION CUBS

CLOUDED LEOPARD CUB

BOBCAT CUB

A mother armadillo gathers leaves to make her burrow cozy for her babies.

Nine-banded armadillo

A coat of armor helps keep this mammal safe.

Despite its name, a nine-banded armadillo can have from seven to 11 bands on its back.

FACTS

KIND OF ANIMAL
mammal

BABY NAME
pup

HOME
warmer forests and grasslands in North and South America

BABIES
four

FOOD
milk, then insects, worms, bird eggs, small animals, fruits, seeds

SIZE AT BIRTH
about the size of a hamster

Nine-banded armadillo pups are born in an underground burrow. The newborn pups have soft, leathery skin. As they get older, it hardens into bony plates that protect the tops of their bodies. Their undersides are soft and covered with hair.

When the pups are a few weeks old, they come out of their burrow. They start looking for food. They use their claws to dig for insects. Their long, sticky tongues can slurp up thousands of ants in one meal!

Armadillo pups are almost always identical quadruplets.

When they're a few months old, male fawns start to grow antlers. Female deer do not have antlers.

Male deer leave their mother after one year, but females usually stay for two.

White-tailed deer

These babies hide on the forest floor.

FACTS

KIND OF ANIMAL
mammal

BABY NAME
fawn

HOME
forests and grasslands in North and South America

BABIES
one or two

FOOD
grass, leaves, bark, twigs

SIZE AT BIRTH
about the size of a bedside table

A fawn stands up as soon as it is born. Its mother licks its body clean. The fawn drinks some milk, then curls into a shallow hole. The spots in its fur make it hard to see. The fawn stays safe while its mom looks for food.

After about a month, Mom leads her fawn out of the forest. They meet up with other mothers and fawns. Sometimes two fawns stand on their hind legs and play-fight with their front legs. They are learning how to protect themselves from wolves and other predators.

Do your clothes help you blend in or stand out?

Sumatran orangutan

This mammal spends its whole life in the trees.

Young orangutans make a lot of noises. They squeak, scream, bark, suck, and burp to communicate.

FACTS

KIND OF ANIMAL
mammal

BABY NAME
infant, baby

HOME
tropical forests in Sumatra, in Southeast Asia

BABIES
usually one at a time

FOOD
milk, then fruits, leaves, bark, nuts, honey, insects, bird eggs

SIZE AT BIRTH
about the size of a football

An orangutan baby is born high in a treetop nest. But the baby and its mother will soon have a new nest.

Each day, the mother finds a different spot to bend and weave branches together. She lines the nest with leaves and makes a pillow with twigs.

Orangutan babies smile at their mothers and cry when they are hungry, just like human babies.

The baby clings to its mother with its hands and feet. It stays close to her for about 10 years.

For the first three months or so, this baby just drinks milk. But once it starts eating fruits and leaves, it can't be a picky eater. Mom gives the baby hundreds of different plants to try!

How many different foods have you eaten today?

CHAPTER 6

Polar Babies

Special fur, fat, and feathers keep these babies warm in the coldest places on Earth.

When they are about two years old, polar bear cubs start living on their own.

Polar bear milk has about 10 times more fat than human milk. The extra fat helps keep cubs warm.

Polar bear

A snowy den is a cozy home for these cubs.

FACTS

KIND OF ANIMAL
mammal

BABY NAME
cub

HOME
far north in Asia, Europe, and North America

BABIES
usually two, up to four at a time

FOOD
milk, then seals, fish, foxes

SIZE AT BIRTH
about as big as a watermelon

A mother polar bear digs a den in the snow. After she gives birth, she hibernates all winter while the cubs drink her milk.

In the spring, the mother and her cubs leave their cozy cave. Their thick fur and a layer of fat called blubber help keep them warm. The cubs play in the snow and slide on the ice. They even get piggyback rides from Mom!

Polar bear cubs take baths by rubbing their fur on the snow.

The cubs watch their mother to learn how to hunt. They eat seals that Mom catches. They dive to catch fish and seals for themselves.

Have you ever built a pretend cave? What was it made of?

The beluga is the only type of whale that can turn its head.

Belugas are sometimes called sea canaries because of all the noises they make. They moo, chirp, whistle, and even make a sound like a bell.

It can take up to eight years for baby belugas to turn white.

BELUGA MOTHER

Beluga whale

Baby belugas are dark gray.

FACTS

KIND OF ANIMAL
mammal

BABY NAME
calf

HOME
far north in the Atlantic and Pacific Oceans

BABIES
one at a time

FOOD
milk, then fish, shrimp, crabs, octopuses

SIZE AT BIRTH
about the length of a park bench

As soon as her calf is born, a mother beluga and another female whale nudge the baby to the water's surface so it can breathe. The calf takes a breath through the blowhole on the top of its head.

The calf will swim next to Mom and drink her milk for about two years. After one year, it learns to catch sea creatures to eat. Belugas have teeth, but they don't use them to chew. They suck food into their mouth and swallow it whole.

The rounded part of a beluga's head is called the melon.

What are some of the different sounds you can make?

King penguin

These parents keep their egg tucked away.

King penguins don't keep their egg in a nest. After the mother lays the egg, the father holds it on his feet. He tucks it under a flap of skin called the brood pouch for about two months. When the penguin chick hatches, it sits in this same spot on either its mother's or father's feet.

Penguins live in large groups called colonies.

The parents take turns going out to sea to find food. While one parent hunts, the other keeps the chick close. When the hunting parent comes back, it sings a special song to find its family in the crowd.

When a king penguin is about a year old, it loses its down and grows waterproof feathers.

EGG

FACTS

KIND OF ANIMAL
bird

BABY NAME
chick

HOME
islands and waters surrounding Antarctica and South America

EGGS
one at a time

FOOD
fish and cephalopods

SIZE AT BIRTH
about the size of a baseball

Early explorers called these chicks "woolly penguins" because their brown down makes them look so different from their parents.

Can you make up a song that would help your family find each other?

FACTS

KIND OF ANIMAL
mammal

BABY NAME
calf

HOME
Arctic Ocean; far north in the Atlantic and Pacific Oceans

BABIES
one at a time

FOOD
milk, then mollusks, worms, birds

SIZE AT BIRTH
about as long as a seven-year-old human is tall

A walrus calf starts growing tusks when it is three to six months old.

Walrus

This mammal's mustache has a job to do.

Walruses live on the ice. A mother walrus keeps close watch over her baby. She takes it with her when she looks for food. Sometimes the calf rides on her back as she swims. Sometimes she holds it in her flippers.

The calf can drink milk anytime it wants—even while its mother is swimming! It follows her to the ocean floor, where it watches her find food. Soon the calf learns how to use the whiskers in its mustache to find clams and snails. It sucks out the meat with its mouth and leaves the shells behind.

WHISKERS

A walrus's blubber can be up to four inches (10 cm) thick. It keeps the walrus warm and helps it float.

A calf stays with its mother for about three years. They live with a herd of other mothers and babies. The walruses in the herd pile up in the sun to stay warm. Female walruses usually join their mother's herd. Males may stay for a few years, but then they leave to find a male herd.

After a month or two, walrus calves molt, or shed their fur. A new coat of fur grows in.

How do you get warm when you're cold?

Calves can eat solid food soon after they're born, but they drink milk for a year.

FACTS

KIND OF ANIMAL
mammal

BABY NAME
calf

HOME
Arctic tundra in Asia, Europe, and North America

BABIES
one at a time

FOOD
milk, then grass, leaves, flowers, roots, seeds, bark

SIZE AT BIRTH
about as big as a carry-on suitcase

Musk ox

These babies have a special hiding place.

Less than an hour after it's born, a musk ox calf can stand and drink milk. In a few hours, it can walk and keep up with its mother. The calf needs Mom to keep it warm while its fur gets thicker. It hides under its mother's skirt, a long layer of fur that protects the calf from the cold.

Young musk oxen play by chasing and pushing one another.

Musk ox calves have nubs where their large, curved horns will grow.

The whole herd works together to keep calves safe. When hungry wolves or other predators come near, adults form a circle around the babies. The young oxen are protected inside the circle.

What's your favorite hiding place?

Arctic wolf

These pups must learn to hunt with their pack.

Arctic wolf pups are born in a den inside a rocky cave. At first, they don't need any food but their mother's milk. After about three weeks, they can see, hear, and walk. It's time for them to leave the den.

FACTS

KIND OF ANIMAL
mammal

BABY NAME
pup

HOME
Arctic tundra in Asia, Europe, and North America

BABIES
usually two to four at a time

FOOD
milk, then arctic hares, caribou

SIZE AT BIRTH
about the size of a grapefruit

Newborn arctic wolf pups are gray. Within a year, their fur will turn white.

Arctic wolves are a type of gray wolf, but their fur stays white year-round. It helps them blend in with the snow.

A wolf pack usually has a mom, a dad, babies, and older brothers and sisters.

All the adults in a wolf pack help care for the pups. They bring meat to new pups. When an adult returns from hunting, the pups whimper and lick around the adult's mouth. The adult throws up food it has chewed and swallowed, and the pups eat it.

When pups play, they practice hunting skills like stalking and pouncing. At about 10 months old, pups start hunting with their pack. When food is hard to find, the leader lets the pups eat first. Over the next year, they will learn to work as a team to hunt large animals like caribou.

How many people are in your family pack?

ANIMAL MAP

You can find where the animals in this book live on this world map.

NORTH AMERICA

American alligator
Arctic wolf
Black-capped chickadee
Black-tailed prairie dog
Brown bear
Burrowing owl
Eastern cottontail rabbit
Golden eagle
Mallard duck
Monarch butterfly
Musk ox
Nine-banded armadillo
Polar bear
Strawberry poison dart frog
White-tailed deer

OCEANS

Beluga whale
Giant Pacific octopus
Olive ridley sea turtle
Sea otter
Walrus
Yellow boxfish

SOUTH AMERICA

Burrowing owl
Eastern cottontail rabbit
King penguin
Nine-banded armadillo
Two-toed sloth
White-tailed deer

EUROPE

Alpine ibex
Arctic wolf
Brown bear
Eurasian badger
Golden eagle
Greater flamingo
Mallard duck
Musk ox
Polar bear

ARCTIC OCEAN

EUROPE

ASIA

AFRICA

PACIFIC OCEAN

AUSTRALIA

OCEANIA

SOUTHERN OCEAN

ANTARCTICA

ASIA
Arctic wolf
Caracal
Brown bear
Eurasian badger
Giant panda
Golden eagle
Greater flamingo
Leopard
Mallard duck
Monarch butterfly
Musk ox
Polar bear
Sumatran orangutan

AFRICA
African elephant
Black-backed jackal
Black rhinoceros
Caracal
Fennec fox
Golden eagle
Greater flamingo
Hippopotamus
Leopard
Mallard duck
Ostrich

AUSTRALIA
Mallard duck
Monarch butterfly
Red kangaroo
White's seahorse

ANTARCTICA
King penguin

PARENT TIPS

Extend your child's experience beyond the pages of this book. A visit to a zoo, aquarium, nature museum, farm, or animal sanctuary may let you see or even touch some of the animals in the book. Keep a list of the animals you see in person. Here are some other activities you can do with National Geographic's *Little Kids First Big Book of Baby Animals.*

ANIMAL NURSERY
(RESPONSIBILITY, EMPATHY)

Help your child set up a nursery for stuffed animals. Talk about things these animals might need if they were real animal babies. What care could humans provide for these babies? What would they need from their real parents?

EGG RACE
(EXERCISE, COORDINATION)

King penguins hold their eggs on their feet (pp. 108–109). Set up a simple racecourse and use a stopwatch to time each other as you move through the course holding a ball on top of your foot or between your ankles. Try it with a kickball or a tennis ball, or any size ball you may have.

NEST BUILDING
(ENGINEERING)

Orangutans build a new nest to sleep in every night (p. 100). Go on a hunt for nest-building materials. Talk about the weight of different twigs, leaves, and grasses. See if the things you find can stick together in a nest shape. Try using mud for "glue" if you like. Or make a human-size nest with blankets and pillows!

WHO'S THAT?

(BIOLOGY)

Use a smartphone to take close-up pictures of the animal furs, skins, and feathers in this book. Show your child the pictures and see if they can identify the animals with these features.

NEST HUNTING

(OBSERVATION)

Early spring is a good time to check trees for birds' nests. Many baby birds hatch at this time of year, and it's easier to see nests in the branches before the leaves fill in completely. Binoculars can help you see from far away. You can also check eaves and gutters and high protected corners on buildings for nests. Don't touch a nest, but if you can look inside, you may see eggs or chicks!

GLOSSARY

BLUBBER: the fat on large animals, such as polar bears, that keeps them warm

BROOD POUCH: a flap of skin where animal parents tuck their eggs or young to develop and grow

BURROW: a hole an animal digs in the ground to use for shelter

CEPHALOPOD: a soft-bodied, legless animal with jaws and limbs, such as an octopus

CORAL REEF: a huge group of coral (soft-bodied animals related to jellyfish) found on the seafloor

CRÈCHE: a large group of young birds cared for by some adults while the other adults in the flock look for food

CROP MILK: a liquid made by certain birds to feed newborn chicks. It comes from a bird's crop, a pouch used to store food inside the throat.

CURRENT: movement of the water in an ocean, lake, or river

DEN: a cave or hole that serves as a home for animals

DOWN: fluffy feathers that baby birds have before their full feathers grow in; also a fluffy layer of feathers that keeps adult birds warm

FLOCK: a group of the same kind of bird that is feeding, resting, or traveling together

HATCHLING: an animal that has just hatched from its egg

HERD: a group of the same kind of animal

HIBERNATE: to spend the winter in an inactive, resting state

MOLT: to shed feathers or skin that will be replaced by new feathers or skin

PACK: a group of the same kind of animal that lives and hunts together

PREDATOR: an animal that hunts for and eats other animals

PREY: an animal that a predator hunts for food

QUADRUPLETS: four babies that are born at the same time

SCUTE: a type of hard scale or plate attached to an animal's skin

TUNDRA: a treeless plain found in the Arctic and in areas just below the Arctic

ADDITIONAL RESOURCES

BOOKS

Delano, Marfé Ferguson. *Explore My World Baby Animals.* National Geographic Kids Books, 2015.

Donohue, Moira Rose. *Little Kids First Big Book of the Rain Forest.* National Geographic Kids Books, 2018.

Drimmer, Stephanie Warren. *Hey, Baby!* National Geographic Kids Books, 2017.

Hughes, Catherine D. *Little Kids First Big Book of Animals.* National Geographic Kids Books, 2011.

Hughes, Catherine D. *Little Kids First Big Book of Birds.* National Geographic Kids Books, 2016.

Nat Geo Wild Animal Atlas. National Geographic Kids Books, 2010.

Spelman, Lucy. *Animal Encyclopedia.* National Geographic Kids Books, 2021.

WEBSITES

A note for parents and teachers: For more information on this topic, you can visit these websites with your young readers.

animaldiversity.org/

kids.nationalgeographic.com/animals/

zooborns.com

INDEX

PHOTO CREDITS

AL: Alamy Stock Photo; GI: Getty Images; MP: Minden Pictures; NG: National Geographic Image Collection; NP: Nature Picture Library; IS: iStockphoto; SS: Shutterstock, **Cover** (duckling), Gregory Johnston/AL; (otters), Alaska Stock/AL; (arctic wolf), Tambako the Jaguar/Moment Open/GI; (alligator), Will E. Davis/SS; (fish), bekirevren/SS; (water behind fish), FlashMovie/SS; (polar bear), Design Pics Inc/NG; (caterpillar), alexbush/Adobe Stock; (lions), Denis-Huot/NP; **back cover** (pandas), Katherine Feng/MP; (prairie dog), Eric Isselee/SS; **spine** (orangutan), Eric Isselee/SS; **Front matter:** 1, Maggy Meyer/SS; 2-3, S. Gerth/age fotostock; 4 (UP), Cathy Keifer/SS; 5 (UP), Tierfotoagentur/AL; 5 (LO), Enjoylife2/IS/GI; **Chapter 1:** 8-9, KAR Photography/AL; 10, Linda Freshwaters Arndt/AL; 11 (UP), Design Pics Inc/AL; 11 (LO), Mhryciw/Dreamstime; 12-13 (RT), Natalia Kuzmina/SS; 12 (LE), Solvin Zankl/NP; 13 (LO), Dirk Ercken/SS; 13 (UP), Chris Rabe/AL; 13 (LE), Huaykwang/SS; 14, Hezi Shohat/SS; 15 (CTR RT), Rabbitti/SS; 15 (LO), Srininivas jadlawad/SS; 15 (UP), Fred Bavendam/MP; **Chapter 2:** 16-17, Pascale Gueret/AL; 18, Klein & Hubert/NP; 19 (UP), Marion Vollborn/BIA/MP; 19 (LO), Marion Vollborn/BIA/MP; 20 (LE), K.A.Willis/SS; 20-21, K.A.Willis/SS; 21 (UP), Craig Dingle/IS/GI; 22 (RT), Eric Isselee/SS; 22 (LO LE), Tony Wear/SS; 22 (UP LE), Auscape International Pty Ltd/AL; 23 (RT), Megan Griffin/SS; 23 (CTR), viktor posnov/Moment RF/GI; 23 (LO LE), Auscape International Pty Ltd/AL; 23 (UP LE), Marc Anderson/AL; 24, Maciej Jaroszewski/IS/GI; 25, Sue Robinson/SS; 26, GoDog Photo/SS; 27, Adri de Visser/MP; 28, Ankevanwyk/Dreamstime; 29, Top-Pics TBK/AL; 30, Frans Lanting/Mint/agefotostock; 31, Mark Newman/FLPA/MP; 32, Glass and Nature/SS; 32 (LE), Breck P. Kent/SS; 33, Sari ONeal/SS; 34 (RT), Jay Ondreicka/SS; 34 (LE), Breck P. Kent/SS; 35 (RT), Geza Farkas/SS; 35 (UP), Michele Zuidema/SS; 36, Anup Shah/Stone RF/GI; 37, Adam Jones/Stone RF/GI; 38, agefotostock/AL; 39, Ivan Kuzmin/AL; **Chapter 3:** 40-41, blickwinkel/AL; 42 (UP), Solvin Zankl/AL; 42, Solvin Zankl/NP; 43, adrian hepworth/AL; 44, Alex Mustard/NP; 45 (RT), Alex Mustard/NP; 45 (LE), Alex Mustard/NP; 46, Kongkham35/SS; 47, Anup Shah/Stone RF/GI; 48 (UP), Wild Wonders of Europe/Allofs/NP; 48 (LE), serkan mutan/SS; 49, Jesus Cobaleda/SS; 50 (LO), Konstantin Novikov/SS; 50, Tony Wu/NP; 51, Fred Bavendam/MP; 52, Richard Mittleman/Gon2Foto/AL; 53 (UP), Design Pics Inc/AL; 53 (LO), Doc White/NP; 54 (CTR), Dennis van de Water/SS; 54 (UP), Melissa Carroll/IS/GI; 54 (LO), Piotr Kamionka/SS; 55, flaviano fabrizi/SS; 56 (UP), Heiko Kiera/SS; 56, C.C. Lockwood/Animals Animals; 57, passion4nature/IS/GI; 58, kwanchai.c/SS; 59, Michel & Christine Denis-Huot/Biosphoto/Science Source; 60, Little Things Abroad/SS; 61, Anup Shah/NP; 62, Stubblefield Photography/SS; 63, Ronny Azran/SS; **Chapter 4:** 64-67, Robert Postma/age fotostock; 66, Erik Mandre/AL; 67, Sergey Uryadnikov/SS; 68, Erik Mandre/SS; 69 (RT), Giedriius/SS; 69 (LE), Giedriius/SS; 70 (LO LE), Naturfoto-Online/AL; 70 (LO), T.Fritz/SS; 70-71 (RT), Ursula Perreten/SS; 71, S. Gerth/age fotostock; 72, Danita Delimont Creative/AL; 73, Michael Callan/FLPA/MP; 74, Michael Forsberg/NG; 74 (UP), W. Perry Conway/Corbis RF Stills/GI; 75, Warren Price Photography/SS; 76, Yossi Eshbol/FLPA/MP; 77, Kim in cherl/Moment RF/GI; 78, All Canada Photos/AL; 79, MTKhaled mahmud/SS; **Chapter 5:** 80-81, George Sanker/NP; 82-83, Kenny Tong/SS; 83, Keren Su/Corbis RF Stills/GI; 83, Katherine Feng/MP; 84, Brian Lasenby/SS; 85, Ross Knowlton Nature Photography/AL; 86, Suzi Eszterhas/MP; 87, Suzi Eszterhas/MP; 88, Ralph Pace/MP; 89, Suzi Eszterhas/MP; 90 (LO), Mark Moffett/MP; 90 (UP), Riccardo Oggioni/AL; 91, Michael and Patricia Fogden/MP; 92 (LO), Suzi Eszterhas/MP; 93 (UP), Trevor Ryan McCall-Peat/SS; 93 (LO), gallas/Adobe Stock; 94 (LE), Eric Isselee/SS; 94 (LO RT), Imagebroker/AL; 95 (RT), Danita Delimont/AL; 95 (UP LE), Dr Ajay Kumar Singh/SS; 95 (LO LE), Tim Fitzharris/MP; 96, Heidi and Hans-Juergen Koch/MP; 97, Bianca Lavies/NG; 98, Imagebroker/AL; 99, Donald M. Jones/MP; 100, Dennis van de Water/SS; 101, Don Mammoser/SS; **Chapter 6:** 102-103, Stone Nature Photography/AL; 104, Robert Sabin/age fotostock; 105 (UP), Robert Harding/AL; 105 (LO), GTW/SS; 106, Andrey Nekrasov/AL; 107, Andrey Nekrasov/imageBROKER RF/GI; 108, JeremyRichards/SS; 109 (UP), Joe McDonald/GI; 110, Paul Nicklen; 111, Paul Nicklen; 112 (UP), Paul Souders/Digital Vision/GI; 112 (LO RT), louise murray/AL; 114, Danita Delimont/AL; 115 (UP), Tsugaru Yutaka/Nature Production/MP; 115 (LO), Kevin Prönnecke/imageBROKER/age fotostock; 116 (LE), Jim Brandenburg/MP; 117 (CTR), Ronan Donovan/NG; 118, Jim Brandenburg/MP; 119, Ronan Donovan/NG; **Back matter:** 122 (UP), Jasmin Merdan/Moment RF/GI; 122 (LO), Frans Lanting/Mint/age fotostock; 123 (UP), Chad Springer/White Door Photo/Image Source/GI; 123 (RT), goodluz/SS; 124 (LE), Andreanita/Dreamstime; 124 (RT), ananth-tp/SS; 125 (LO), mike powles/Stone RF/GI; 125 (UP), Amy Lutz/SS; 128, giedriius/Adobe Stock

For my not-so-little babies, Julia, Allie, and Caroline — M. M.

The publisher acknowledges and thanks zoologist Lucy Spelman and herpetologist Robert Powell for their expert review of this book. Many thanks also to project manager Grace Hill Smith, researcher Michelle Harris, and photo editor Annette Kiesow for their invaluable help with this book.

Since 1888, the National Geographic Society has funded more than 14,000 research, conservation, education, and storytelling projects around the world. National Geographic Partners distributes a portion of the funds it receives from your purchase to National Geographic Society to support programs including the conservation of animals and their habitats. To learn more, visit natgeo.com/info.

For more information, visit nationalgeographic.com, call 1-877-873-6846, or write to the following address:

National Geographic Partners, LLC
1145 17th Street N.W.
Washington, D.C. 20036-4688 U.S.A.

For librarians and teachers: nationalgeographic.com/books/librarians-and-educators

More for kids from National Geographic: natgeokids.com

For rights or permissions inquiries, please contact National Geographic Books Subsidiary Rights:
bookrights@natgeo.com

Designed by Nicole Lazarus, Design Superette

Library of Congress Cataloging-in-Publication Data
Names: Myers, Maya, author.
Title: Little kids first big book of baby animals / Maya Myers.
Description: Washington, D.C. : National Geographic Kids, 2022. | Series: Little kids first big book | Includes index. | Audience: Ages 4-8 | Audience: Grades 2-3
Identifiers: LCCN 2020005716 (print) | LCCN 2020005717 (ebook) | ISBN 9781426371462 (hardcover) | ISBN 9781426371479 (library binding) | ISBN 9781426371486 (ebook)
Subjects: LCSH: Animals--Infancy--Juvenile literature.
Classification: LCC QL763 .M97 2022 (print) | LCC QL763 (ebook) | DDC 591.3/92--dc23
LC record available at https://lccn.loc.gov/2020005716
LC ebook record available at https://lccn.loc.gov/2020005717

Printed in China
21/PPS/1

EURASIAN BROWN BEARS